Focus on . . . Charts and Graphs age 7 – 8

Written by Michael Tonge
Illustrated by James Robins

About this book

Focus on . . . Charts and Graphs gives practice in specific numeracy skills, as described in the guidelines for the National Numeracy Strategy. The book is in line with the National Curriculum of England and Wales and the 5–14 programme in Scotland.

Instructions are written simply so that children can work independently. This helps children's confidence, while giving valuable extra practice in important skills.

Answers are given on pages 30–32, so that either you or your child can mark the work.

A wipe-clean page at the end of the book will give your child even more practice at understanding charts and graphs.

How to help your child

Let your child decide how long he or she wants to work on the book. Give praise for every effort that has been made. Give your child a reward medal for each page.

If your child has problems with a page or a type of activity, note the description of skills and discuss it with the class teacher.

How to use the stickers and progress chart

You can use the stickers to reward good work and to look for areas that need extra practice.

- Ask your child to use the gold, silver or bronze stickers to show how much he or she enjoyed each page.
- Put a matching sticker on the Progress Chart – this will help to show where your child may need some extra support.

About the author

Michael Tonge is a headteacher and an experienced mathematics teacher.

Commissioning Editor: Shirley Jackson
Editor: Rachel Murawa
Designer: The Camel Corps
Produced by: Bender Richardson White

Published in Great Britain by
Egmont World Limited,
Deanway Technology Centre,
Wilmslow Road, Handforth,
Cheshire SK9 3FB

Printed in Italy

ISBN 0 7498 4616 X

Frequency charts

Look at the picture and then fill in the chart at the bottom of the page.

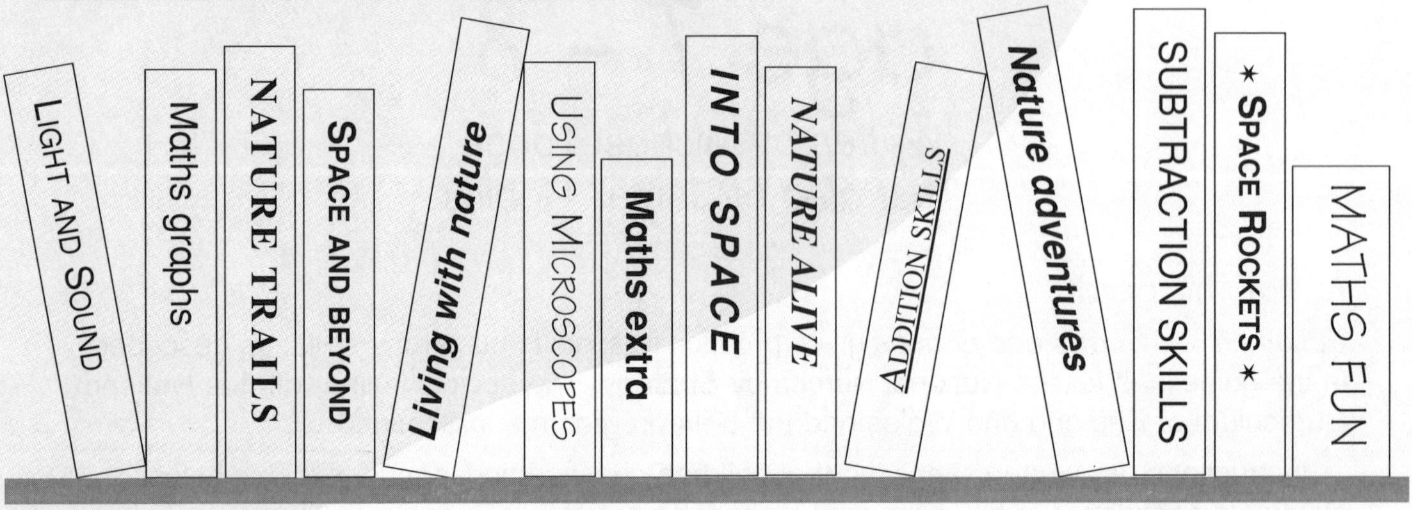

Books	Number of books	Number of books on loan
Nature		
Space		
Maths		
Science		

Which section has the highest number of books? ____________________

What is the difference between the highest number and the lowest number of books? ____________________

Each book section should have 10 books.
Calculate how many books are out on loan for each section.

MEDAL WINNER!

We call the number of times something happens 'frequency'.

Read this short passage. Then fill in the frequency chart.
Count all the words with two letters, three letters and so on.

We all need maths for life. We need to be able to add up, to solve problems and to work out money for ourselves and for others. Maths is one of the most important things we will ever learn but it is not always easy and sometimes we have to work very hard indeed at it. We sometimes think we are not doing very well at all and that we are always stuck. We have to think, though: if we never find anything hard and we never get stuck then we never really learn anything.

Number of letters in word	Frequency
2	
3	
4	
5	
6	
7	
8	
9	
10	

What kind of word has the highest frequency?

____

Are there any word-types that do not appear?

____

What is the difference between the highest frequency and the next to the highest frequency?

____

Frequency charts and block graphs

Fill in the answers on this addition square.

+	1	2	3	4	5	6	7	8	9	10
1	2									
2		4	5	6	7	8	9	10	11	12
3			6	7	8	9	10	11	12	13
4				8	9	10	11	12	13	14
5	6				10	11				
6	7					12	13			
7	8						14	15		
8	9							16	17	
9	10								18	19
10	11	12	13	14	15	16	17	18	19	20

How frequently does each group of answers appear?

Answer	Frequency
1–4	
5–8	
9–12	
13–16	
17–20	

Which answers have the highest frequency? ____________________

Why do you think these answers appear the most? ____________________

Which answers have the lowest frequency? ____________________

Why do you think these answers appear least? ____________________

Which number does not appear at all in the answers? ____________________

Why does this number not appear at all? ____________________

MEDAL WINNER!

This block graph shows the number of television sets sold at a shop in one week.

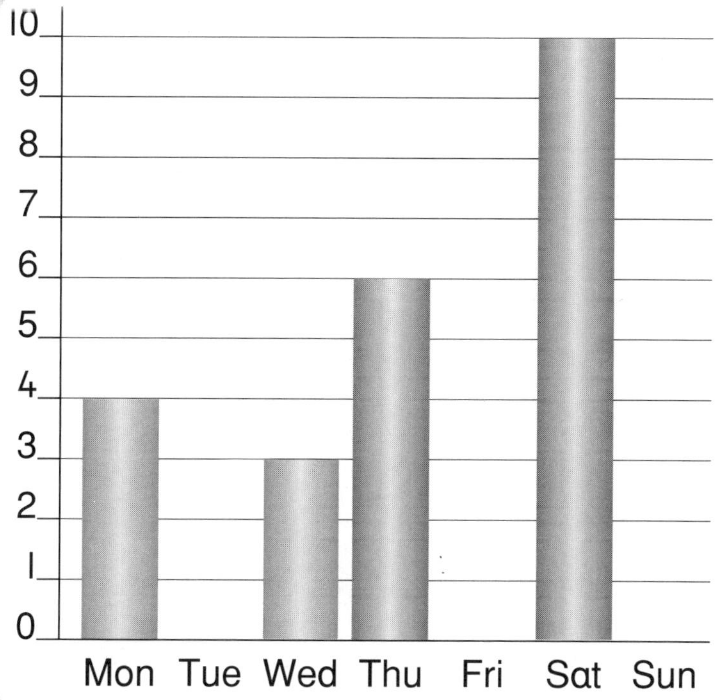

On Tuesday they sold half as many televisions as on Thursday.
Fill in the number of televisions sold on Tuesday.

On Friday they sold twice as many televisions as on Wednesday.
Fill in the graph to show how many televisions they sold on Friday.

On which day did they sell the most televisions? ________________

What was the highest number of televisions sold? ________________

On which day do you think the shop was closed? ________________

Explain your answer. ________________________________

The shop began the week with 50 televisions.
How many televisions do they still have left? ________________

Block graphs and tally charts

These two block graphs show two weeks of group visits to a museum.

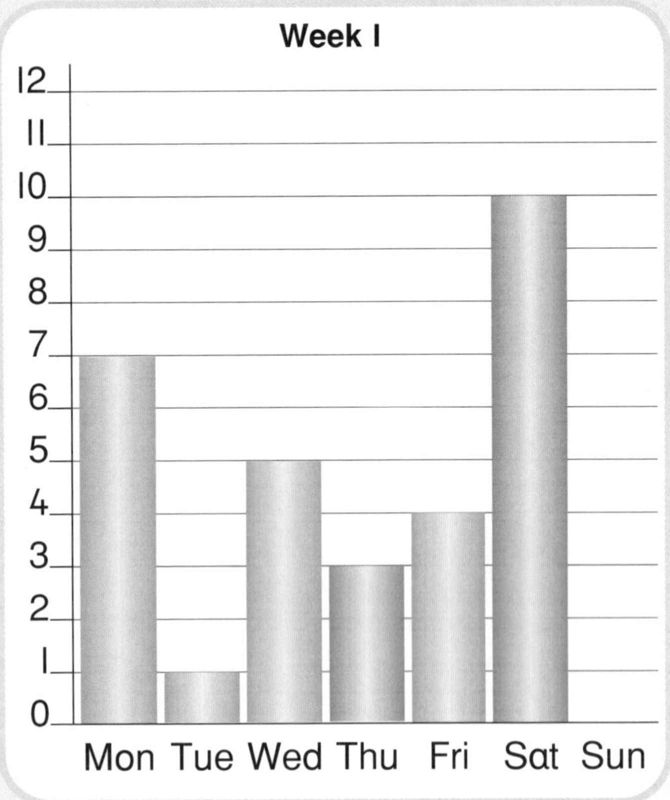

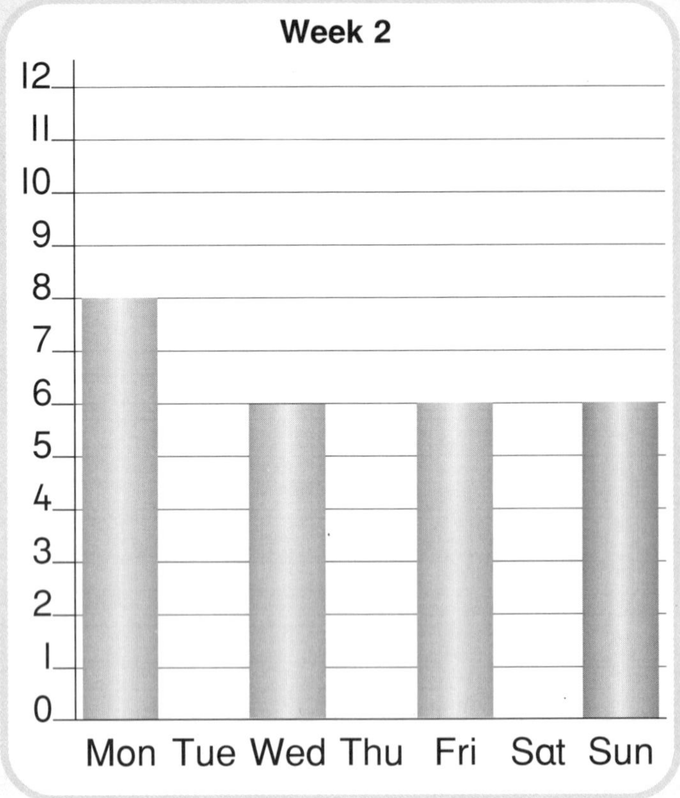

Use the information below to finish off the graphs:

Twice as many groups visited the museum on the second Thursday than on the first Thursday.

One and a half times as many groups visited the museum on the first Sunday than on the second Sunday.

On the second Saturday, twice as many groups visited than on the day before.

Now answer these questions:

Which week had the most number of group visits? ____________________

What day, for both weeks, was the busiest day? ____________________

How many groups visited the museum over the two weekends? __________

The museum has to close for a new exhibit to be put up.
What day would you close the museum on? ____________________

Explain your answer. ____________________

____

____

MEDAL WINNER!

A tally chart is a way of recording information. For every item or event we put a tally mark I. When we reach a tally of five we put a line through them like this ~~IIII~~.

How many does each of these tally marks represent?

II ____________________

III ____________________

~~IIII~~ ____________________

IIII ____________________

~~IIII~~ ~~IIII~~ II ____________________

Look at this tally chart showing the number of cars sold at a garage during one month.

Cars	Number of cars
Red	~~IIII~~ IIII
Blue	~~IIII~~ ~~IIII~~ ~~IIII~~
Yellow	~~IIII~~ ~~IIII~~ II
Black	~~IIII~~ ~~IIII~~ II
White	~~IIII~~ III

What was the most popular colour of car for that month? ____________________

What was the difference between the most popular and least popular car colours? ____________________

What two colours sold the same amount of cars? ____________________

Why is it important when selling cars to know what the most and least popular colours are? ____________________

Tally charts and block graphs

Look at this picture of a traffic jam. Use the tally chart below to make a record of all the different types of vehicles in the jam.

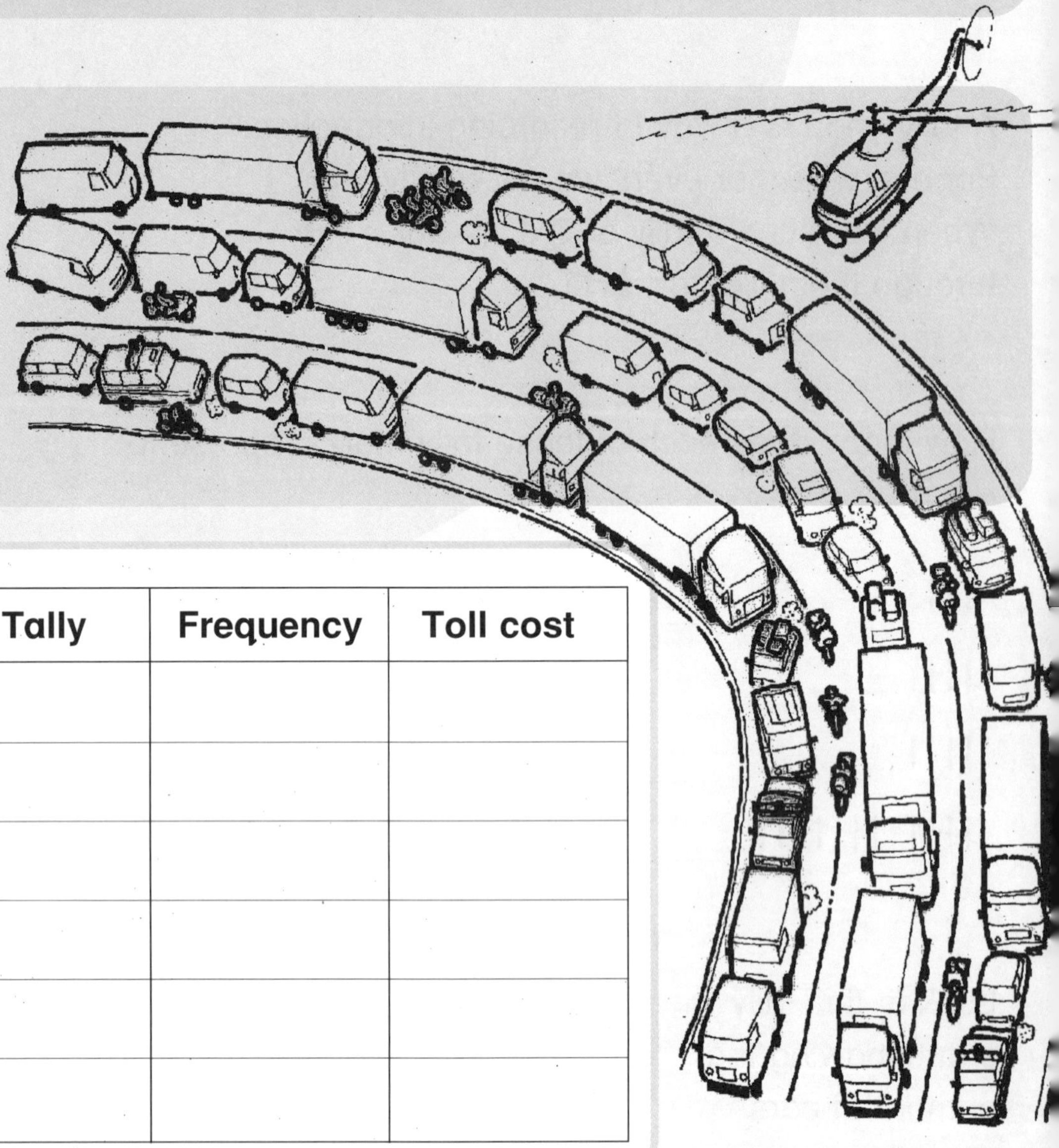

Vehicle	Tally	Frequency	Toll cost
Cars			
Motorbikes			
Vans			
Lorries			
Police cars			
Helicopters			

Write each vehicle's tally as a number in the frequency column.

Which vehicle has the highest frequency?

____

What is the difference between the number of lorries and motorbikes?

____

Look at this sign for tunnel prices.

Lorry £2
Cars £1
Motorbikes £0.50
Vans £1.50
Police Vehicles £0.20

Calculate how much money each set of vehicles had to pay to use the tunnel. Write it on the chart.
Which vehicle does not have to pay any tolls?

____

MEDAL WINNER!

Use the information on the tally chart below to draw a block graph.
The first one is done for you.

Day	Boxes of crisps sold
Mon	𝍸 𝍸 II
Tue	𝍸 IIII
Wed	𝍸 𝍸 I
Thu	𝍸 II
Fri	
Sat	𝍸 𝍸 𝍸
Sun	𝍸 II

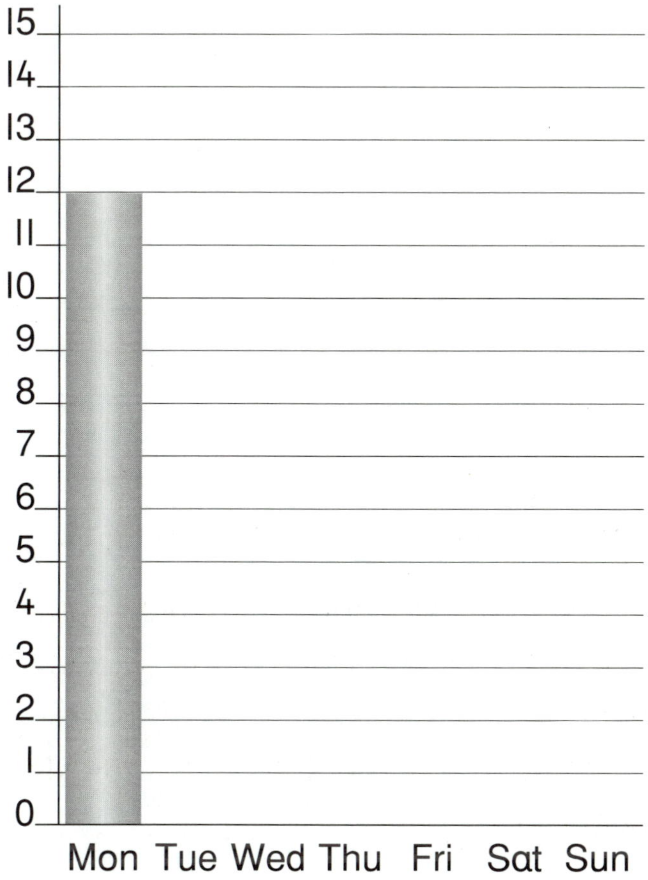

Which day is the crisp shop closed? ______________________

Explain your answer. ______________________

When does the crisp shop sell the most number of crisps? ______________________

On which chart is it easier to see the best sales day? ______________________

Explain why. ______________________

____

____

Bar charts

This is a bar chart for the number of hours a family spent watching television over a two-week holiday.

Week 1

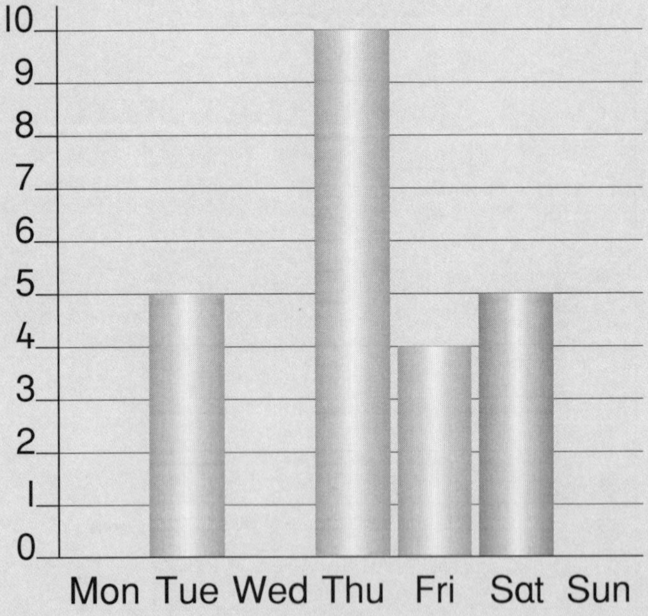

Week 2

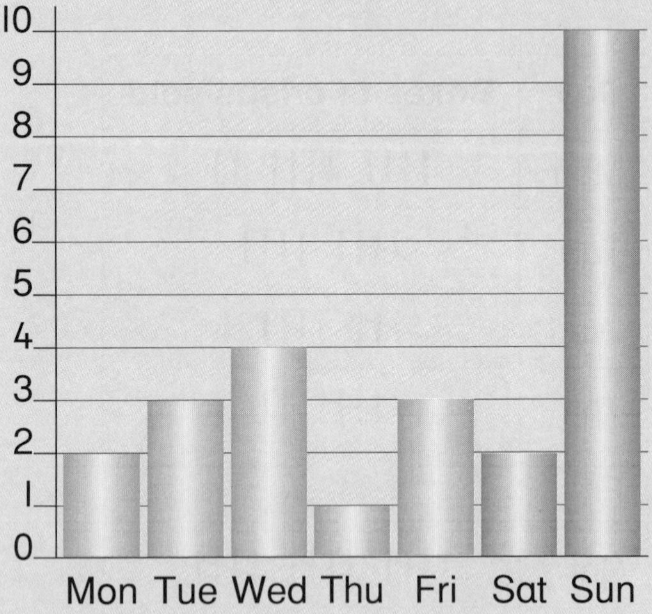

Complete the bar chart using this information.

The family watched twice as much television on the first Monday than on the second.

The family watched half as much television on the first Wednesday than on the second.

Look at the bar chart and then answer these questions.

Two days of the holidays had very bad weather. On which days do you think the weather was worst?

Explain why you think this.

____

____

____

____

____

What day during the holiday do you think the family went out all day?

Explain why you think this.

____

____

____

Which week did the family spend the most amount of time watching television? ______________________________

MEDAL WINNER!

Use the information in this chart to complete the bar chart.

Time	Cars Washed
8:00–9:00	9
9:00–10:00	4
10:00–11:00	3
11:00–12:00	8
12:00–1:00	6
1:00–2:00	3
2:00–3:00	2
3:00–4:00	3
4:00–5:00	1
5:00–6:00	10

Give your chart a title. Name each axis.

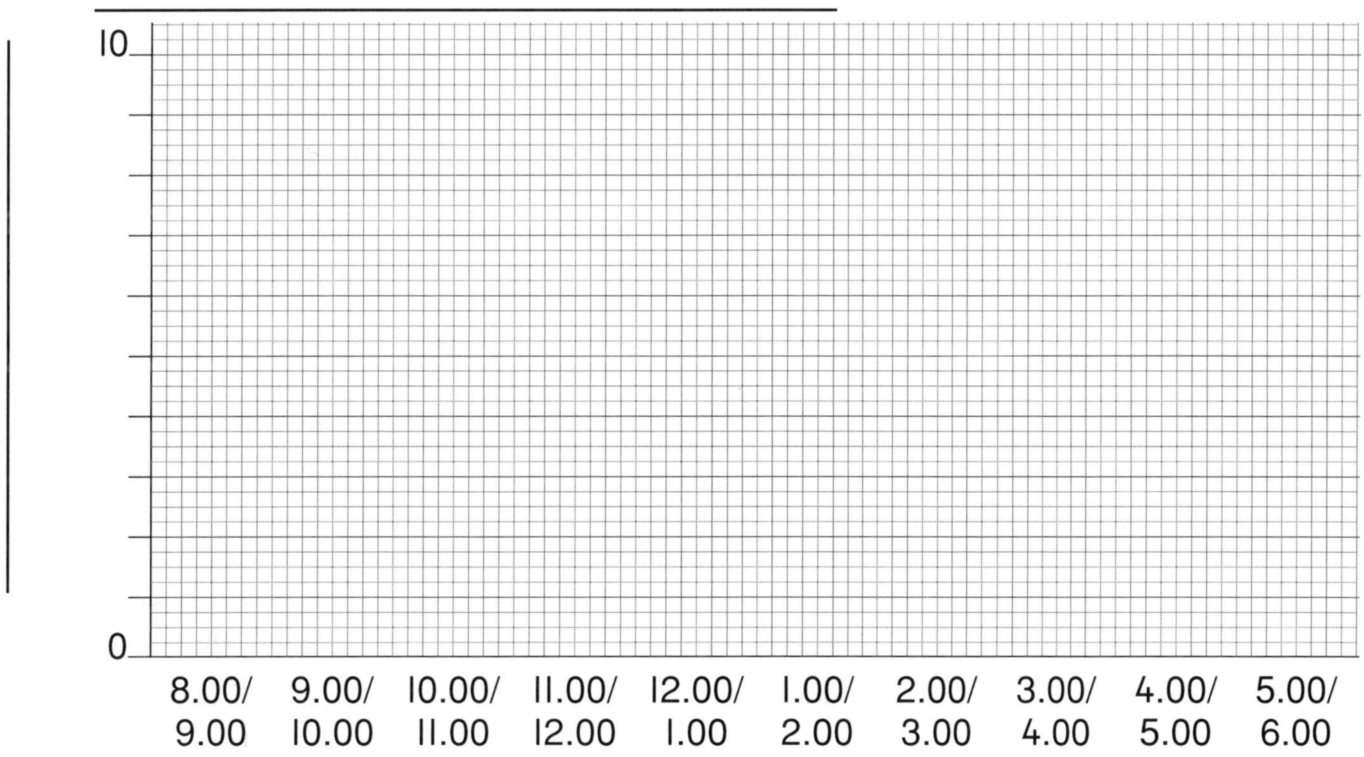

Tip:
The title of a chart tells you what the chart is about.
Each axis measures something. On this chart the axis at the side measures how many cars; the axis along the bottom measures time.

Bar charts and pictograms

Use the information in the chart below to complete the bar chart.

Time	Buses
8:00–9:00	10
9:00–10:00	5
10:00–11:00	3
11:00–12:00	2
12:00–1:00	6
1:00–2:00	4
2:00–3:00	3
3:00–4:00	2
4:00–5:00	3
5:00–6:00	10

Give your chart a title. Name each axis.

____

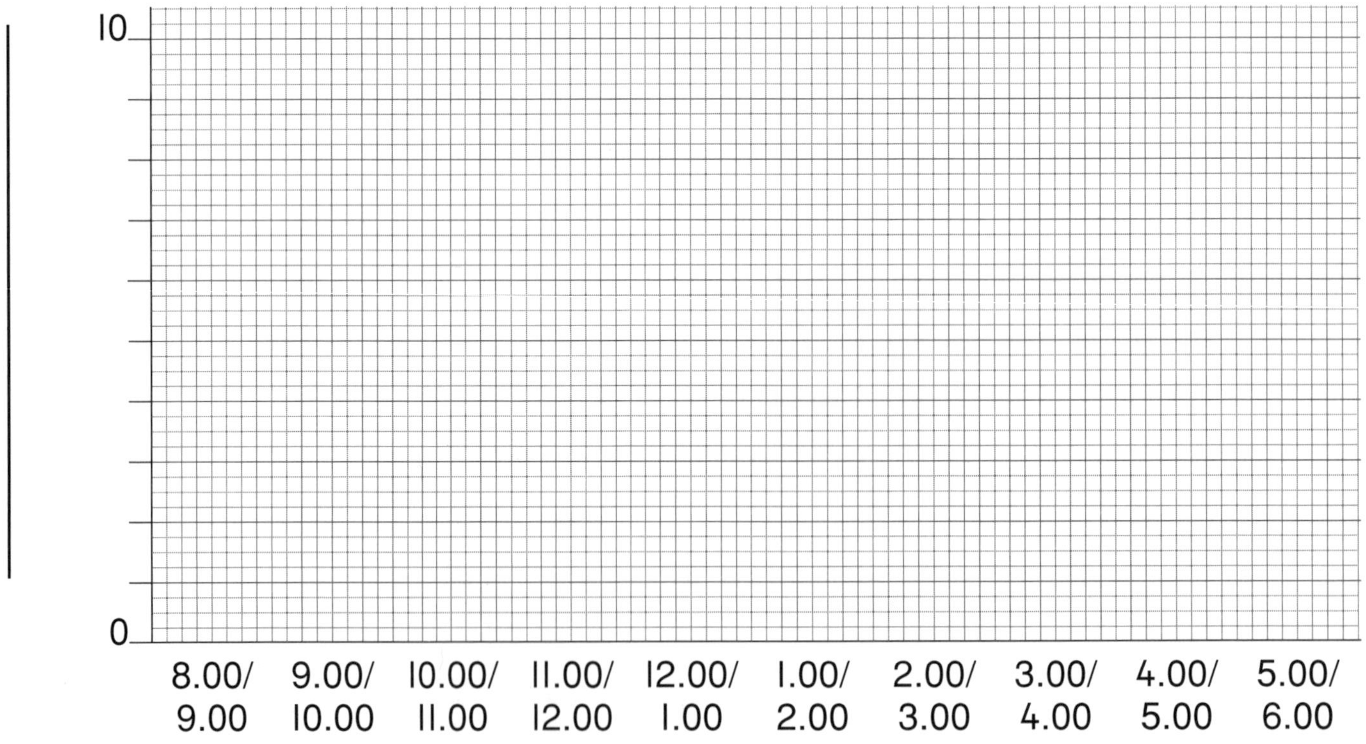

____

When does the greatest number of buses travel? ____________________

Why do you think there are more buses at these times? ____________________

MEDAL WINNER!

A pictogram is a way of showing information.
It uses pictures to represent things.
This picture  means 1 person.

How many people do you think this picture shows? ____________________

Look at this pictogram. Write down next to the pictogram the number of people who go swimming for each day.

Days	Swimming	How many?
Mon		
Tue		
Wed		
Thu		
Fri		
Sat		
Sun		

What day did the greatest number of children go swimming? ____________________

What day did half that number go swimming? ____________________

What day did the least number of children go swimming? ____________________

Pictograms

Sometimes a pictogram is used to stand for a number of people, things or events.

This cake  stands for 10 cakes.

How may cakes does each of these pictograms stand for?

 ____________________

 ____________________

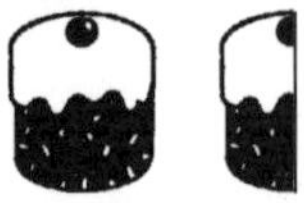 ____________________

This car  means that 10 cars have been sold.
How many cars did this company sell each month?

Month	Sales	How many?
Jan		
Feb		
Mar		
Apr		
May		
Jun		
Jul		
Aug		
Sep		
Oct		
Nov		
Dec		

Tip:
Half a car = 5 cars sold.

MEDAL WINNER!

This pictogram stands for the number of hot school meals eaten in a year.

 = 20 plates

Here is a list of all the meals eaten each month. Use the symbol above to show the information as a pictogram.

	Meals Eaten	Pictogram
January	𝍸 𝍸 𝍸 𝍸 𝍸 𝍸 𝍸 𝍸 𝍸	
February	𝍸 𝍸 𝍸 𝍸 𝍸 𝍸 𝍸	
March	𝍸 𝍸 𝍸 𝍸 𝍸 𝍸	
April	𝍸 𝍸 𝍸 𝍸 𝍸	
May	𝍸 𝍸 𝍸 𝍸	
June	𝍸 𝍸 𝍸 𝍸	
July	𝍸 𝍸	
August		
September	𝍸 𝍸 𝍸 𝍸 𝍸 𝍸 𝍸 𝍸	
October	𝍸 𝍸 𝍸 𝍸 𝍸 𝍸 𝍸 𝍸 𝍸 𝍸	
November	𝍸 𝍸 𝍸 𝍸 𝍸 𝍸 𝍸 𝍸 𝍸 𝍸 𝍸	
December	𝍸 𝍸 𝍸 𝍸 𝍸 𝍸 𝍸 𝍸 𝍸 𝍸 𝍸 𝍸 𝍸	

Why does the number go up during December? ______________________

What month do you think school is closed? ______________________

Explain your answer. ______________________

Which way of showing the information is easiest to understand? ______________________

Explain your answer. ______________________

____

Pictograms

This book  represents 20 books.

How many books do these pictures represent?

 ____________  ____________ 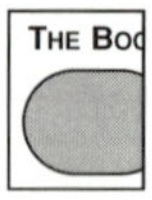 ____________

Name	Books	How many?
Razia	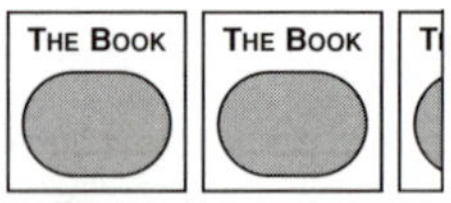	____________
Melanie		____________
Sheena		____________
Jerry	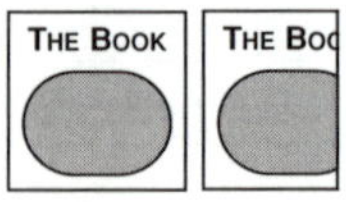	____________
Peter	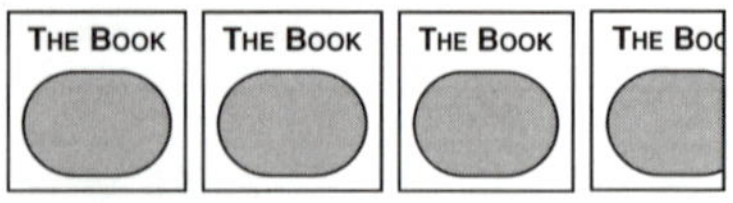	____________
Simon		____________
Jameel		____________
Michael		____________
Jordan		____________

Complete the chart to say how many books each child has read in a year.

Who has read the most books?

____

Who has read the least number of books.

____

Draw a pictogram showing the difference between the highest and lowest number of books read.

____

MEDAL WINNER!

This picture of a wheel stands for 9 stolen bikes. Draw the pictogram for the following numbers of stolen bikes using the same picture.

18 stolen bikes

27 stolen bikes

12 stolen bikes

15 stolen bikes

3 stolen bikes

6 stolen bikes

Design your own symbol for 10 fireworks sold.

Use your symbol to show:

5 fireworks

20 fireworks

15 fireworks

35 fireworks

Graph scales

Look at this graph.

If you look at the graph you can see the numbers on the side go up in jumps or intervals of 5. Why do you think that is?

____

____

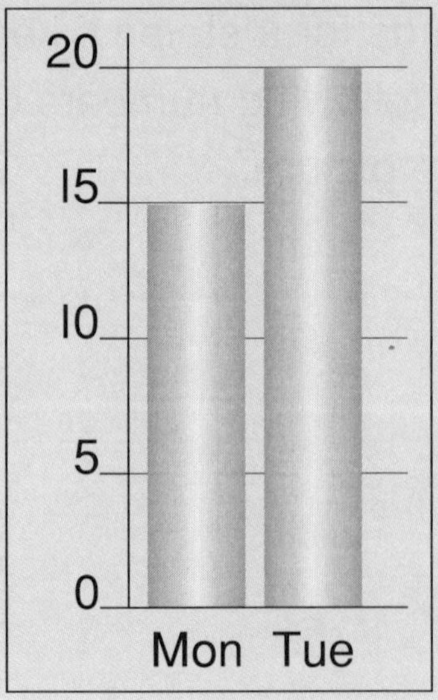

Now look at this graph.

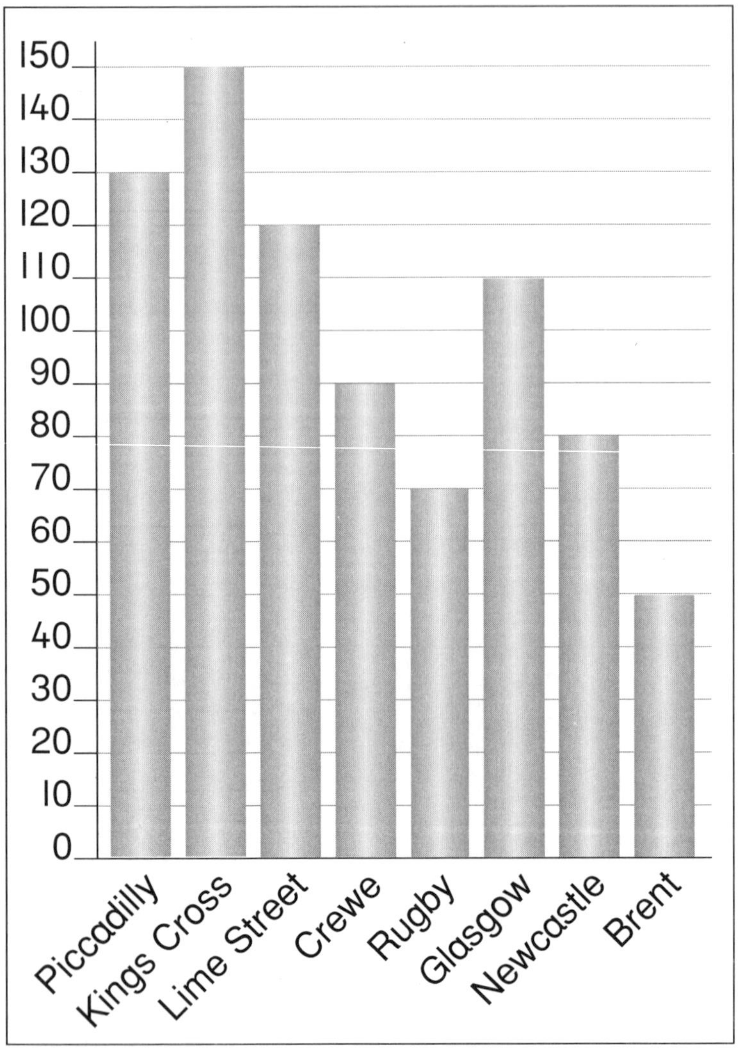

It shows the number of trains going into different stations each day. What steps (intervals) does the side axis go up in?

____

Why do you think it goes up by this amount?

____

Which station has the most number of trains?

____

Which station has the least number of trains?

____

How can you tell this by looking at the graph?

____

MEDAL WINNER!

Look at this graph.

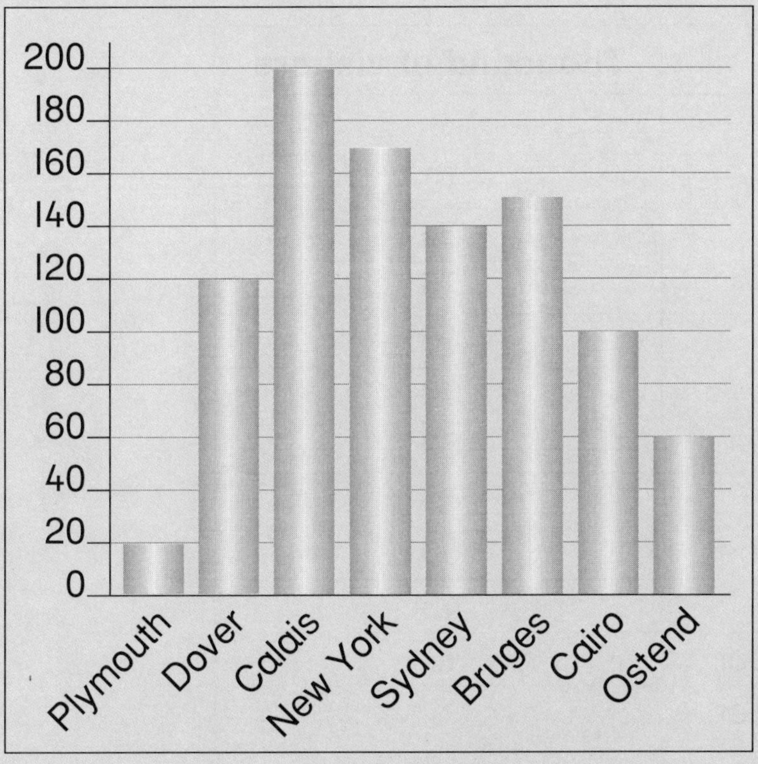

How much do the intervals increase by on the side axis?

____

Which port has the highest number of ships?

____

Which port has the lowest number of ships?

____

You can count the difference between the number of intervals for Dover and Calais like this. Place a ruler across the lower one, Dover, and then count up in steps of twenty until you reach the highest one, Calais.

What is the difference between the number of ships arriving in Dover and Calais? ____________________

In the same way work out the difference between the number of ships arriving in:

New York and Sydney ____________ Cairo and Dover ____________

Plymouth and New York ____________ Ostend and Calais ____________

Ostend and Bruges ____________

Drawing bar graph scales

Use this frequency chart to draw a bar chart.

What is the best increase for each interval?

____

The interval is how much the numbers along the side go up by!

Theme park visitors for this year	
Month	**Thousand of visitors**
January	5
February	10
March	20
April	35
May	45
June	60
July	60
August	60
September	50
October	45
November	10
December	5

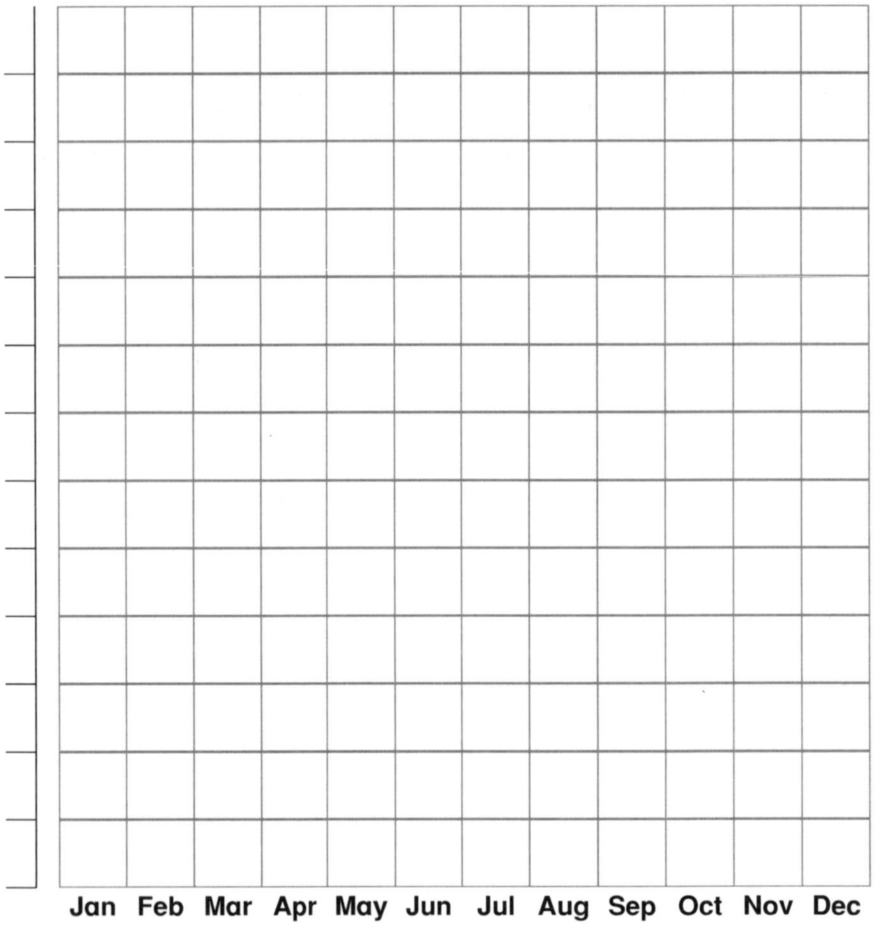

MEDAL WINNER!

Look at the information below. It shows the number of computers sold during a year.

	Hundreds
Jan	33
Feb	29
Mar	22
Apr	17
May	12
Jun	14
Jul	19
Aug	23
Sep	24
Oct	28
Nov	30
Dec	39

Look at the computer sales carefully and the graph template shown below. Decide on the best interval to show all the information. It may help if you think of the numbers as being just below or just above the nearest multiple of 5. (For example, 23 is nearly 25!)

____

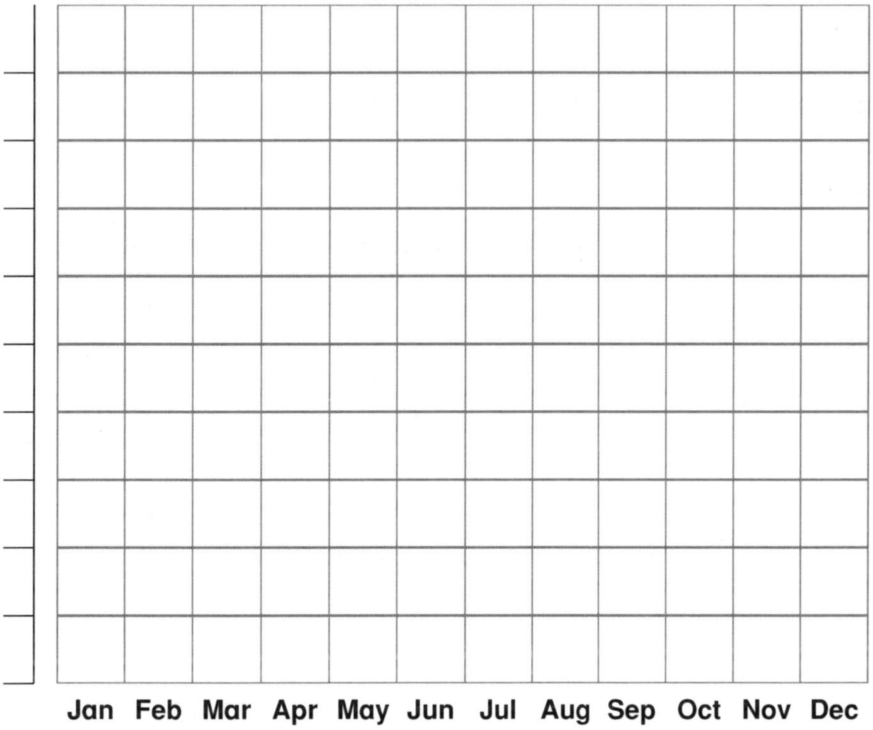

Why do you think they sold more computers during December and January?

____

____

____

Time bar charts

Look at the information shown in this graph.

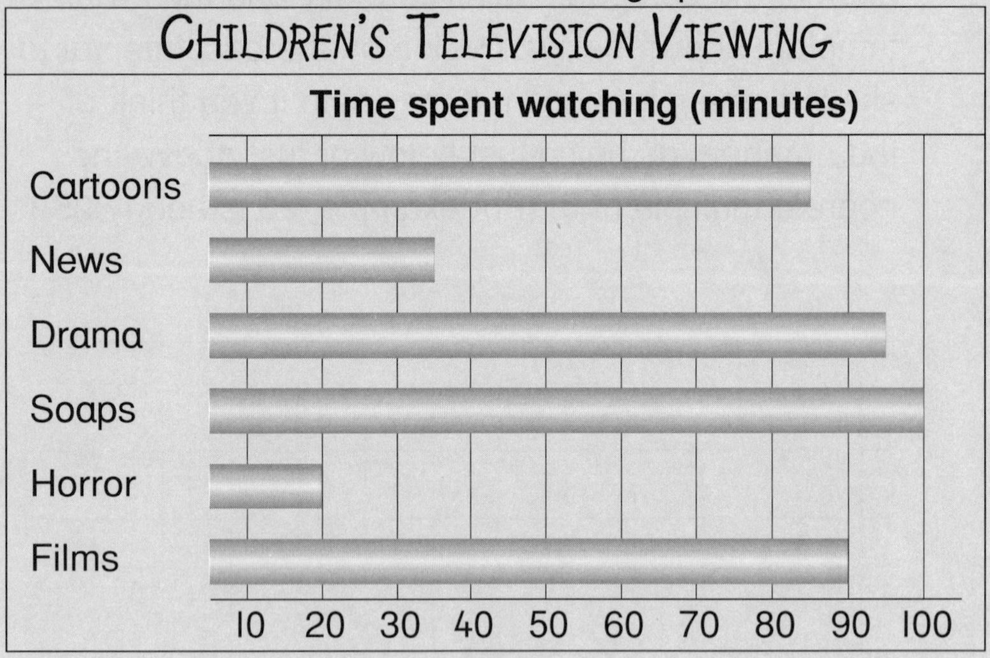

What did children spend most of their time watching?

How many programmes were watched for longer than an hour?

What is the least watched type of programme?

Which type of programme is watched for $1\frac{1}{2}$ hours?

Which type of programme is watched for 1 hour longer than the news?

If you wanted to put an advert on during a programme, which programme would you choose? Explain your answer.

MEDAL WINNER!

Look at the cooking times shown on this chart.

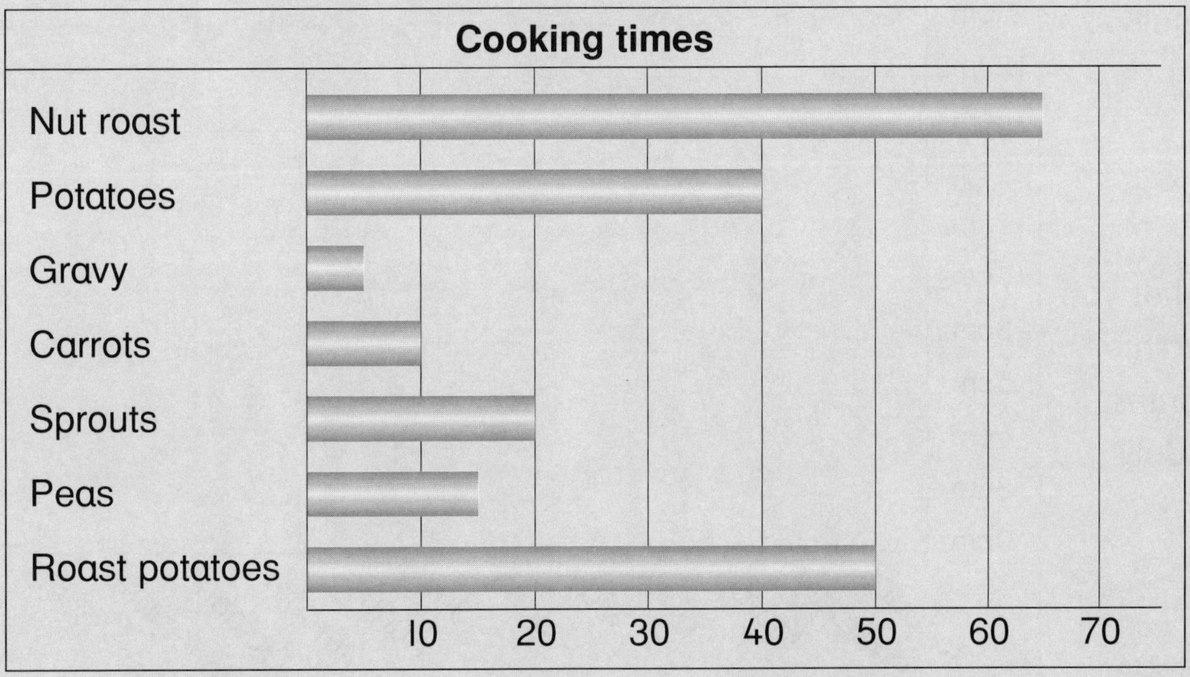

Which food takes the longest to prepare? ______________________

Imagine you had to prepare this meal for your family.

If the meal had to be ready at 1:00 in the afternoon what times would you have to start cooking each piece of food to make sure they were all ready at 1:00 o'clock?

Nut roast

Potatoes

Gravy

Carrots

Sprouts

Peas

Roast potatoes

Bar chart trends

Look at this bar chart carefully.

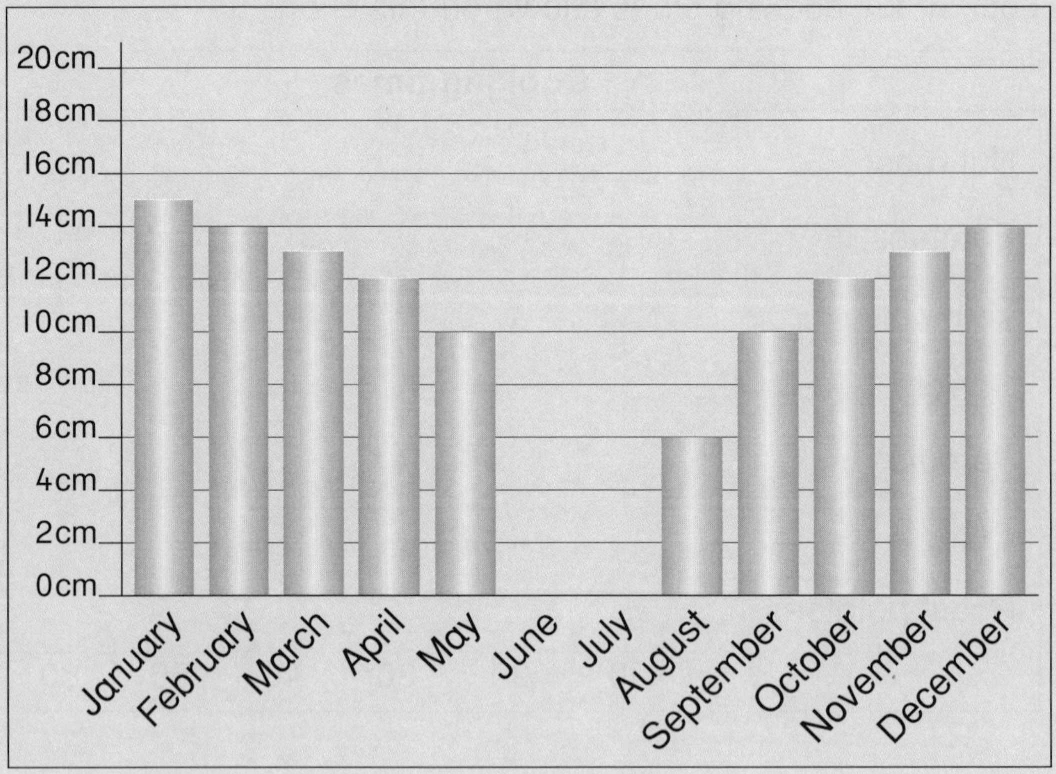

Draw in what you think the rainfall for the months June and July might be.

Tip:
You may want to draw a line joining up the tops of each of the bars to help you.

Which month had the highest rainfall? ______________________

Which months had the lowest rainfall? ______________________

Why do you think the children who recorded this information got these results?

____

____

____

MEDAL WINNER!

Look at this bar chart carefully.

What do you think the temperatures might be for the missing months? Draw them in.

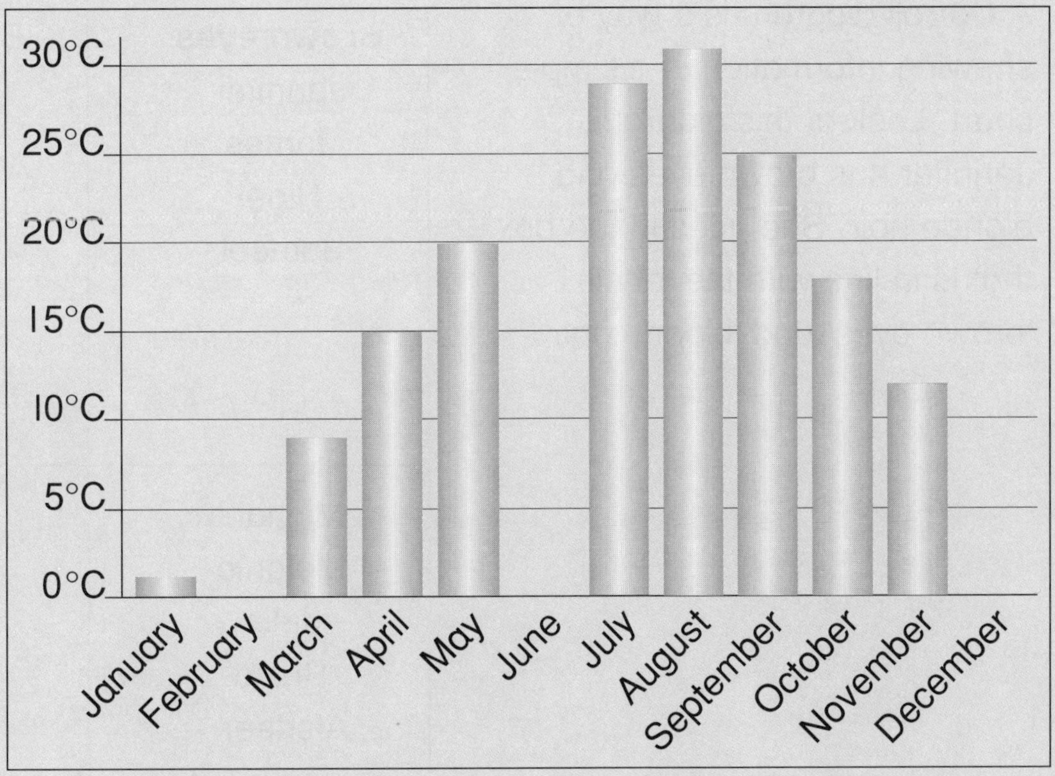

Tip:
You may want to draw a line joining up the tops of each of the bars to help you.

Which month had the highest temperature? ____________________

Which months had the lowest temperature? ____________________

What was the difference in temperature between the months of August and February? ____________________

Which months are colder than 15°C? ____________________

____

Why do you think the children who recorded this information got these results?

____

____

Carroll diagrams

A Carroll diagram is a way of showing information in a simple chart. Look at this diagram. Jennifer has brown eyes and blonde hair. She goes in the box that is in line with the labels 'brown eyes' and 'blonde hair'.

Brown eyes	Blue eyes	
Jennifer James Nigel Jameel	Janet Peter Sheila Jordan	**Blonde hair**
Michael Melanie Julie Razia Afsheen	Abigail Frank Harry James	**Brown hair**

In this Carroll diagram how many children have brown hair and blue eyes?

What are the names of the children who have brown hair and brown eyes?

Put these children in the correct sections of the diagram.

Jill	blue eyes	brown hair
Guy	brown eyes	blonde hair
Philip	brown eyes	brown hair
Cath	blonde hair	blue eyes
Jamie	brown eyes	blonde hair.

MEDAL WINNER!

Look carefully at the information shown here.

Write down the name of each object in the correct box of the Carroll diagram.

	Over £5	Under £5
Round		
Not round		

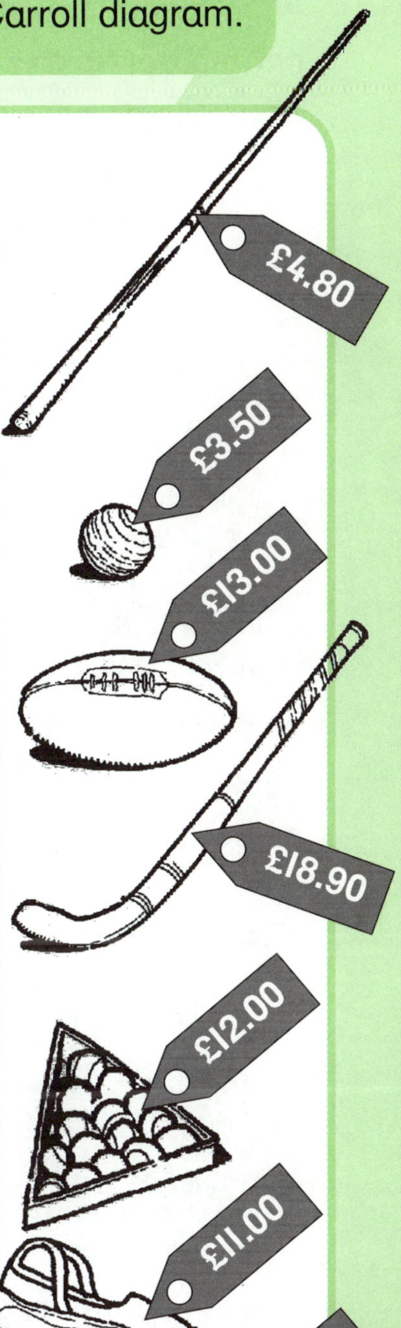

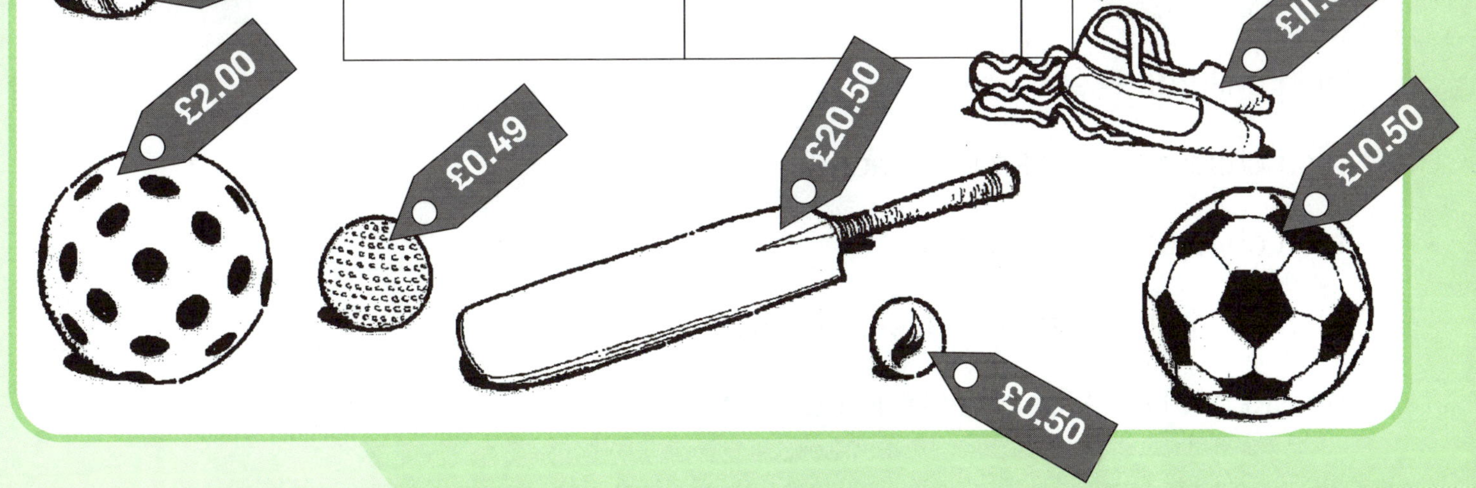

Venn diagrams

A Venn diagram is a way of showing information in a simple chart. Look at this Venn diagram.

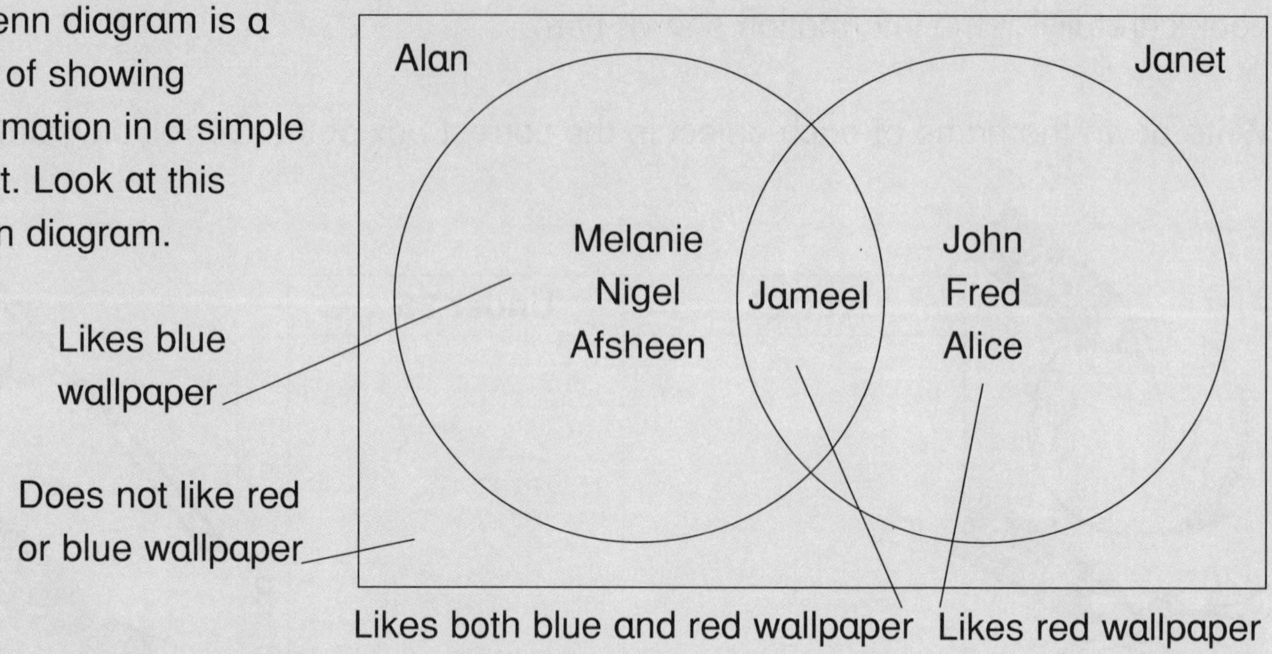

Look at this Venn diagram. Answer the questions below it.

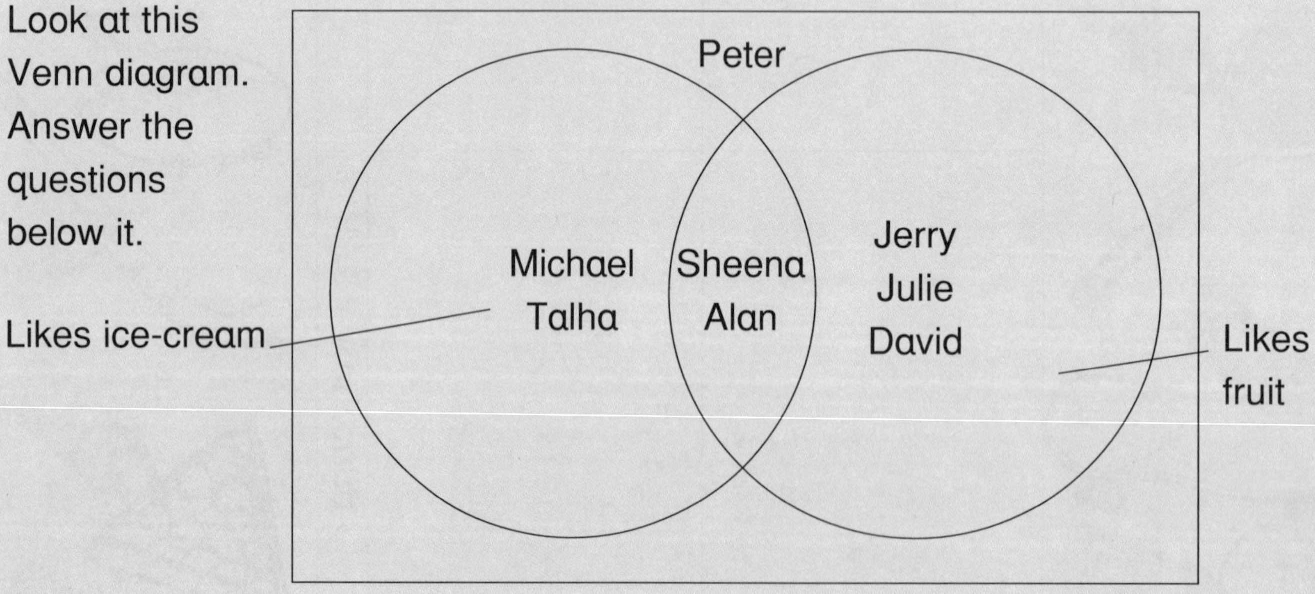

Which children like ice-cream? ______________________

Which children only like ice-cream? ______________________

Which children only like fruit? ______________________

Which children like fruit and ice-cream? ______________________

Who did not like either fruit or ice-cream? ______________________

MEDAL WINNER!

Use the information on this tally chart to complete the Venn diagram.

Name	Pets		
	Dog	Cat	None
Melanie		✔	
Alex	✔	✔	
Jordan		✔	
Agibail	✔		
Michael	✔		
Janet	✔		
Peter		✔	
Judith	✔	✔	
Asif			✔
Sheena	✔		
Lesley	✔		

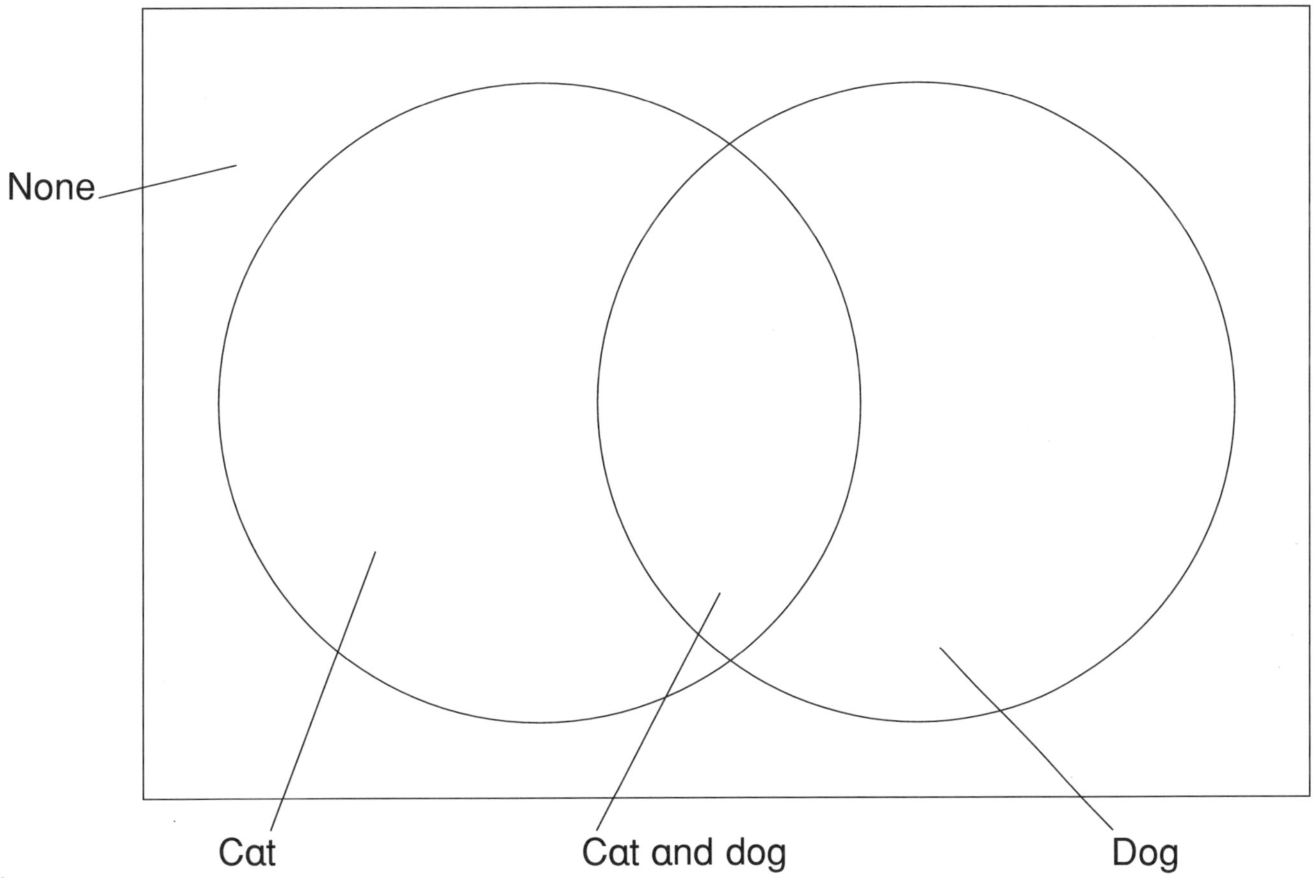

Answers

Page 2

Books	No of books	No of books on loan
Nature	4	6
Space	3	7
Maths	5	5
Science	2	8

maths; 3

Page 3

No of letters in word	Frequency
2	27
3	20
4	20
5	14
6	7
7	0
8	3
9	4
10	0

highest frequency: 2–letter words
7–letter and 10–letter words do not appear
difference between highest and next highest frequency is 7

Page 4

+	**1**	**2**	**3**	**4**	**5**	**6**	**7**	**8**	**9**	**10**
1	2	**3**	**4**	**5**	**6**	**7**	**8**	**9**	**10**	**11**
2	**3**	4	5	6	7	8	9	10	11	12
3	**4**	**5**	6	7	8	9	10	11	12	13
4	**5**	**6**	**7**	8	9	10	11	12	13	14
5	6	**7**	**8**	**9**	10	11	**12**	**13**	**14**	**15**
6	7	**8**	**9**	**10**	**11**	12	13	**14**	**15**	**16**
7	8	**9**	**10**	**11**	**12**	**13**	14	15	**16**	**17**
8	9	**10**	**11**	**12**	**13**	**14**	**15**	16	17	**18**
9	10	**11**	**12**	**13**	**14**	**15**	**16**	**17**	18	19
10	11	12	13	14	15	16	17	18	19	20

Answer	Frequency
1–4	6
5–8	22
9–12	36
13–16	26
17–20	10

9–12 because all the numbers are part of a number bond that makes 9–12;
1–4 because they are small numbers, there are not many number bonds that make 1–4;
1 does not appear in the answers because the lowest addition is 1 + 1 = 2

Page 5

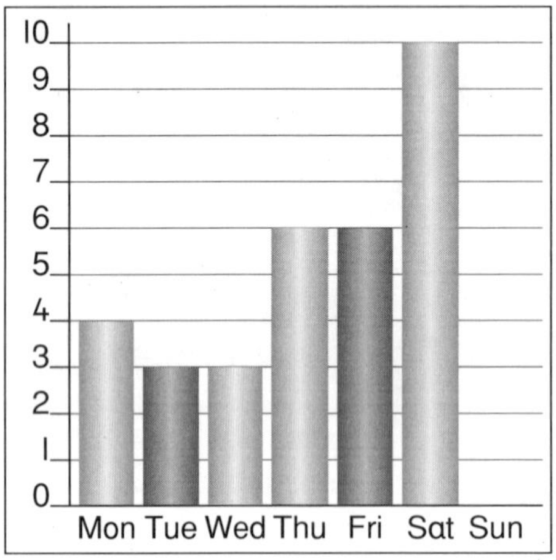

Saturday; 10; Sunday because no TVs were sold; 18

Page 6

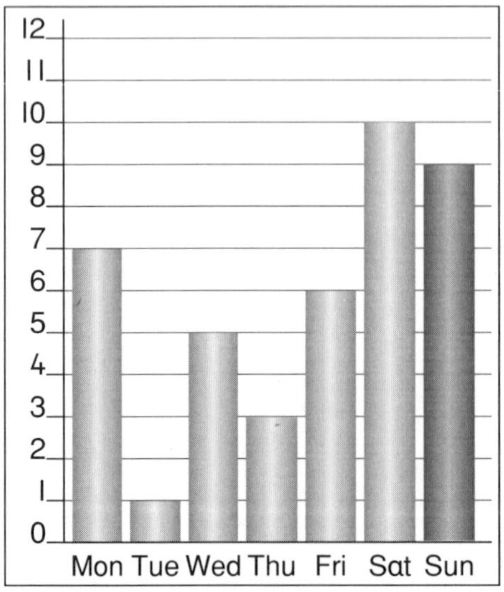

week 2; Saturday; 37; Tuesday because it has the fewest visitors

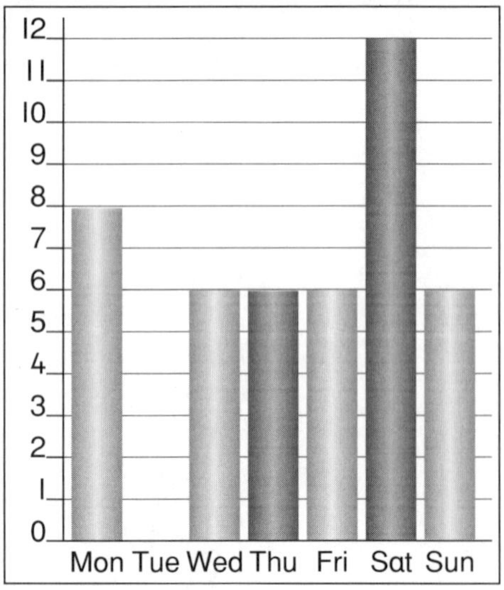

Page 7

2; 3; 5; 4; 12
blue; 7; yellow, black; so you know what to order

Page 8

Vehicle	Tally	Frequency	Toll co
Cars	𝍸 𝍸 IIII	14	£14.00
Motorbikes	𝍸 𝍸 I	11	£5.50
Vans	𝍸 IIII	9	£13.50
Lorries	𝍸 III	8	£16.00
Police cars	III	3	60p
Helicopters	I	1	0

cars; 3; helicopters

Page 9

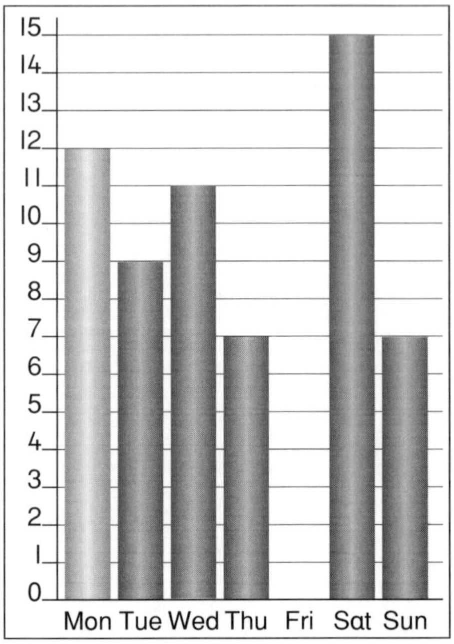

Friday because there were no sales; Saturday; most people prefer a block graph because you can see at a glance, but accept a good explanation for either option.

Page 10

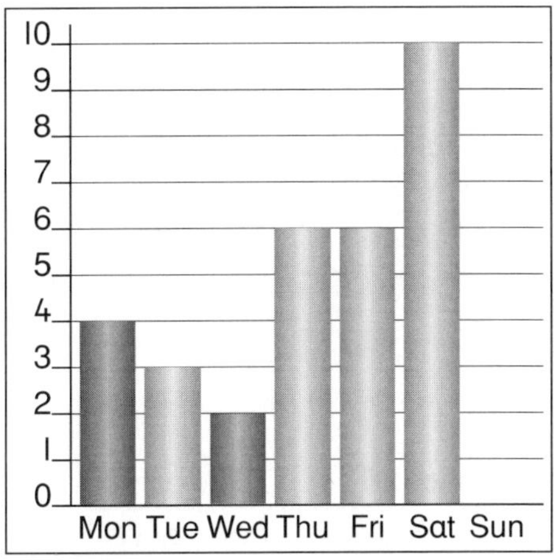

The first Thursday and the second Sunday because they watched lots of TV. The family went out on the first Sunday because no TV watched.
Watched most TV in week 1.

Page 11

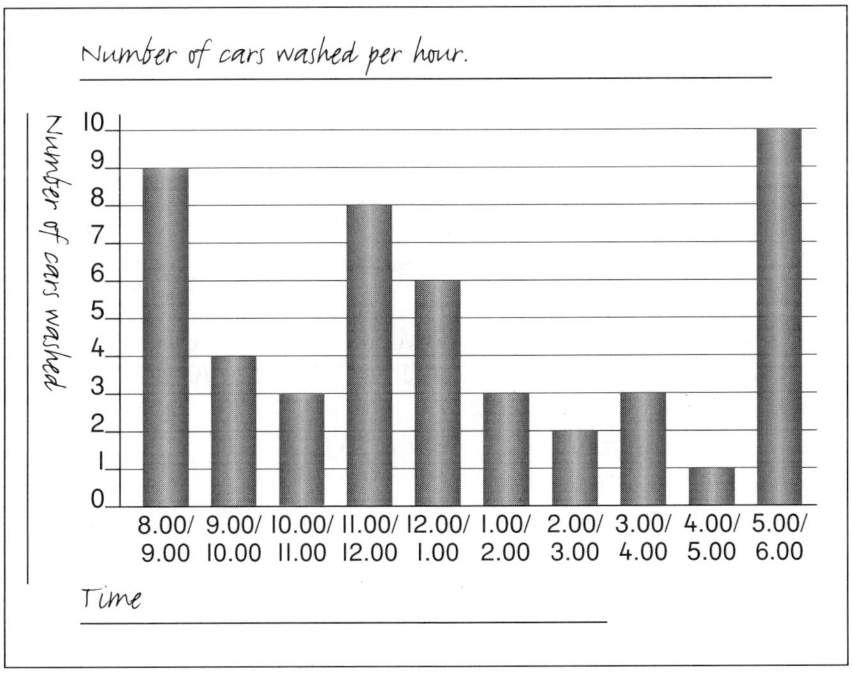

Page 12

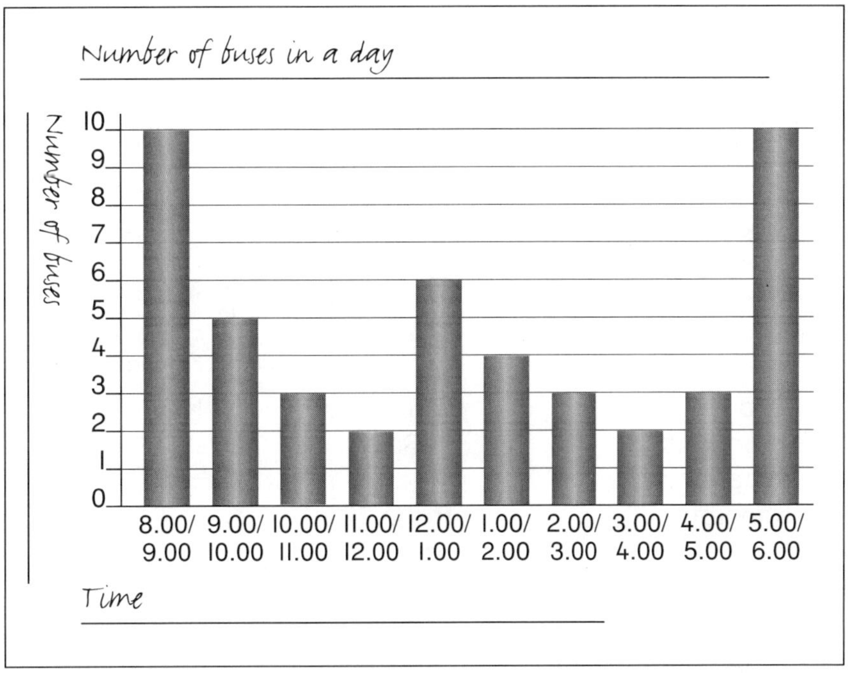

morning and evening; it is when people are travelling to and from school and work

Page 13

9; Mon = 9, Tues = 12; Wed = 7; Thu = 6; Fri = 5; Sat = 8; Sun = 10
Tuesday; Thursday; Friday

Page 14

30; 5; 15
Jan = 30; Feb = 20; Mar = 65; Apr = 10; May = 20; Jun = 5; Jul = 35; Aug = 55; Sep = 45; Oct = 15; Nov = 5; Dec = 25

Page 15

January = $2\frac{1}{4}$ pictograms;
February = $1\frac{3}{4}$ pictograms;
March = $1\frac{1}{2}$ pictograms;
April = $1\frac{1}{4}$ pictograms;
May = 1 pictogram;
June = 1 pictogram;
July = $\frac{1}{2}$ pictogram;
August = 0;
September = 2 pictograms;
October = $2\frac{1}{2}$ pictograms;
November = $2\frac{3}{4}$ pictograms;
December = $3\frac{1}{4}$ pictograms
The weather is colder; August because no hot dinners were sold; pictograms can be clearer because there is less counting, but accept a good explanation for either

Answers continued

Page 16
$\frac{1}{2}$ pictogram = 10; $\frac{1}{4}$ pictogram = 5; $\frac{3}{4}$ pictogram = 15
Razia: 45; Melanie: 65; Sheena: 55; Jerry: 35; Peter: 75; Simon: 25; Jameel: 60; Michael: 55; Jordan: 60
Peter has read the most books. Simon has read the least; You should draw $2\frac{1}{2}$ pictograms

Page 17
18 = 2 pictograms;
27 = 3 pictograms;
12 = $1\frac{1}{3}$ pictograms;
15 = $1\frac{2}{3}$ pictograms;
3 = $\frac{1}{3}$ pictogram;
6 = $\frac{2}{3}$ pictogram
5 = $\frac{1}{2}$ pictogram;
20 = 2 pictograms;
15 = $1\frac{1}{2}$ pictograms;
35 = $3\frac{1}{2}$ pictograms

Page 18
To fit more information into a smaller space
10; to fit the information in; Kings Cross; Brent; by comparing the height of the bars

Page 19
20; Calais; Plymouth
80; 30; 150; 90; 20; 140

Page 20
intervals of 5 or 10

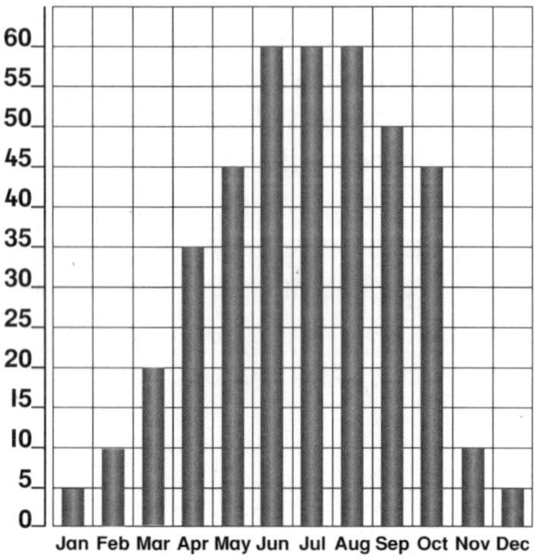

Page 21

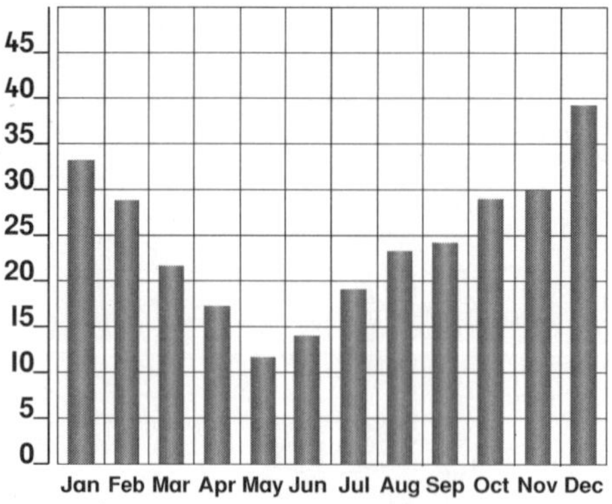

Christmas and the New Year Sales

Page 22
soaps; 4: cartoons, dramas, soaps, films; horror; films; dramas; soaps because they have the most viewers

Page 23
nut roast; nut roast – 11:55; potatoes – 12:20; gravy – 12:55; carrots – 12:50; sprouts – 12:40; peas – 12:45; roast potatoes – 12:10

Page 24
June and July should be between 10 cm and 6 cm.
January; July and August; they have summer between June and August when it is drier, and winter from October to March when it is wetter.

Page 25
February should be 5–7°C, June between 22–27°C, December between 2–6°C
August; January, December; 25°C; January, February, November and December because it is winter during those months.

Page 26
4; Michael, Melanie, Julie, Razia, Afsheen

Brown eyes	Blue eyes	
Jennifer James Nigel Jameel **Guy** **Jamie**	Janet Peter Sheila Jordan **Cath**	**Blonde hair**
Michael Melanie Julie Razia Afsheen **Philip**	Abigail Frank Harry James **Jill**	**Brown hair**

Page 27

Over £5	Under £5	
pool balls football cricket ball	bouncy ball spotty ball marble golf ball stripey ball	**Round**
football boot tennis racquet skateboard rugby ball hockey stick ballet shoes cricket bat	snooker cue	**Not round**

Page 28
Michael, Talha, Sheena, Alan; Michael, Talha; Jerry, Julie, David; Sheena, Alan; Peter

Page 29

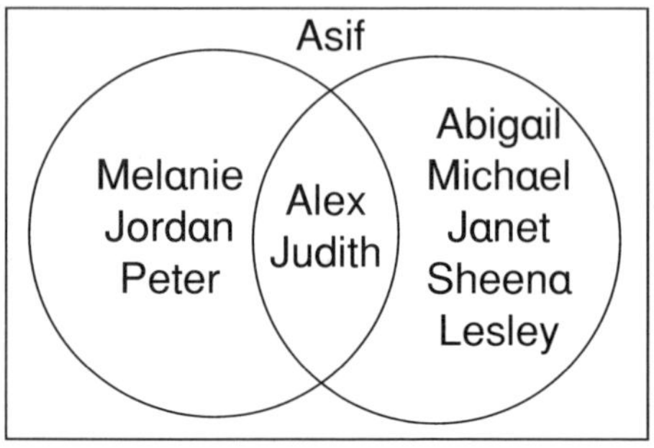